W9-CDF-206

Crayfish

by Phyllis W. Grimm
photographs by Jerry Boucher

Lerner Publications Company • Minneapolis, Minnesota

To my grandchildren
—PWG

To my number one grandson, Zachary Alan Boucher
—JB

The publisher wishes to thank Dr. Keith A. Crandall, Curator of Crustacea, M. L. Bean Museum, Brigham Young University, for his careful review of this manuscript. Special thanks to our young helper, Kyle Crandall.

Thanks to our series consultant, Sharyn Fenwick, elementary science and math specialist. Mrs. Fenwick was the winner of the National Science Teachers Association 1991 Distinguished Teaching Award. She also was the recipient of the Presidential Award for Excellence in Math and Science Teaching, representing the state of Minnesota at the elementary level in 1992.

Additional photographs are reproduced with permission from: © Richard Thom/Visuals Unlimited, p. 8; © Gary Meszaros/Visuals Unlimited, p. 9; © Todd Walsh, p. 11; © Bill Beatty/Visuals Unlimited, p. 12; © Maslowski/Visuals Unlimited, p. 25; © Breck P. Kent, pp. 28, 29, 33, 34; © Philip Gould/Corbis, pp. 40, 41.

Early Bird Nature Books were conceptualized by Ruth Berman and designed by Steve Foley. Series editor is Joelle Riley.

Lerner Publications Company
A division of Lerner Publishing Group
241 First Avenue North
Minneapolis, Minnesota 55401 U.S.A.

Website address: www.lernerbooks.com

Library of Congress Cataloging-in-Publication Data

Grimm, Phyllis W.
    Crayfish / by Phyllis W. Grimm ; photographs by Jerry Boucher.
        p. cm. — (Early bird nature books)
        Includes index.
        Summary: Describes the physical characteristics, behaviors such as the search for food and eating habits, method of reproduction, habitat, and survival challenges of this group of crustaceans.
        ISBN-13: 978–0–8225–3030–5 (lib. bdg. : alk. paper)
        ISBN-10: 0–8225–3030–9 (lib. bdg. : alk. paper)
        1. Crayfish—Juvenile literature. [1. Crayfish.] I. Boucher, Jerry, 1941– II. Title. III. Series.
    QL444.M33 G76   2001
    595.3'84—dc21   99-050716

Manufactured in the United States of America
3 4 5 6 7 8 – JR – 10 09 08 07 06 05

Contents

Be a Word Detective

Can you find these words as you read about the crayfish's life? Be a detective and try to figure out what they mean. You can turn to the glossary on page 46 for help.

abdomen	**crustaceans**	**instars**
antennas	**exoskeleton**	**molting**
cephalothorax	**gills**	**oxygen**
compound eyes	**in berry**	**swimmerets**

There is more than one way to form plurals of some words. The word antenna has two possible plural endings—either an e or an s. In this book, s is used when many antennas are being discussed.

Chapter 1

Crayfish are also called crawfish or crawdads. What animals are related to crayfish?

Not a Fish!

 Crayfish are fun to watch. They scoot around at the edges of lakes and in streams. But be careful! If you pick one up, you may get pinched by one of its big claws.

Crayfish are not fish! Crayfish belong to a group of animals called crustaceans (cruhs-TAY-shuhnz). All crustaceans have a hard outer shell. Most crustaceans live in water. There are many species, or kinds, of crustaceans. They live all over the world. Lobsters, crabs, and shrimp are also crustaceans.

The coral banded shrimp is a crustacean. There are over 30,000 species of crustaceans.

Scientists have found over 500 species of crayfish. Crayfish live in almost every part of the world.

A few species of crayfish live in caves. They are pale and blind.

Some species of crayfish dig holes with high walls to live in. The walls are called chimneys.

Most crayfish live in freshwater. Usually they live in shallow parts of lakes, ponds, rivers, streams, and swamps. Some crayfish live in wet fields. They dig holes to live in.

A fish's gills are just behind its eyes.

Crayfish are not fish. But they breathe like fish. Crayfish breathe with gills. Gills take a gas called oxygen (AHK-sih-juhn) out of the water. Most creatures need oxygen to live. You get oxygen from the air you breathe into your lungs.

Most crayfish are about 3 inches long. This is about as long as a crayon. Some crayfish are only 1 inch long. That's about the size of a small paper clip. The largest crayfish species is about 16 inches long. That's as long as your arm!

The scientific name of the world's largest crayfish is Astacopsis gouldi. *It lives in the country of Australia.*

Crayfish come in many different colors.
They can be pink, white, red, black, green, or other
colors. Many crayfish are more than one color.

*Some crayfish have bright color on just a few parts
of their body.*

Over half of all crayfish species live in the United States.

Most species of crayfish live 2 or 3 years. Large species of crayfish live longer than that. Some large crayfish in the southern United States live 6 to 8 years. Very large crayfish may live 20 years.

Chapter 2

This crayfish's scientific name is Orconectes virilis. This species is common in the United States. How many main body parts does a crayfish have?

Lots of Parts

A crayfish's body has two main parts. One part is called the cephalothorax (seh-fuh-luh-THOR-ax). The cephalothorax is the crayfish's head and chest. It is the front part of the crayfish. The second part of the crayfish's body is called the abdomen (AB-duh-muhn). The abdomen is the back part of the crayfish's body.

A crayfish's legs are attached to its cephalothorax.

A crayfish uses its eyes to see all around it.

Crayfish have two eyes on their cephalothorax. The eyes are on short stalks. They can move around while the rest of the crayfish stays still. Crayfish can grow new eyes if their eyes are hurt.

A crayfish's eyes are compound eyes. Compound eyes are made up of many small eyes. Compound eyes help a crayfish to see moving objects.

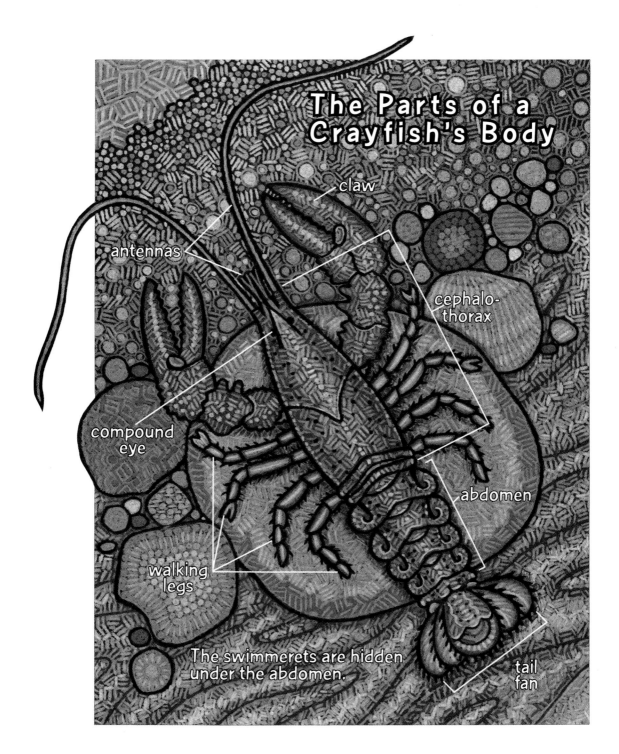

The Parts of a Crayfish's Body

claw

antennas

cephalo-thorax

compound eye

abdomen

walking legs

The swimmerets are hidden under the abdomen.

tail fan

Crayfish have four antennas (an-TEH-nuhz). Crayfish use their antennas to feel, taste, and smell. Two of the antennas are short. The other two antennas are long.

A crayfish's antennas have tiny hairs all over them. The hairs help the crayfish to feel, taste, and smell.

Crayfish have 12 tiny mouthparts on their cephalothorax. The mouthparts help crayfish taste and smell. A crayfish uses some of its mouthparts to break up food and to put food into its mouth. Some mouthparts move water over the crayfish's gills so it can breathe.

Crayfish can grow new antennas if their old ones are hurt.

A crayfish has two big claws. It uses the claws to grab and hold objects. The crayfish also uses its claws to protect itself from enemies.

Male crayfish have larger claws than female crayfish have.

A crayfish uses its walking legs to crawl around.

Crayfish have eight walking legs. They are under the cephalothorax. Crayfish use their walking legs to move forward, backward, and sideways.

A crayfish's tail fan helps it to move quickly.

Crayfish have 10 swimmerets under their abdomen. Swimmerets look like small legs. But crayfish don't use them to walk.

The end of a crayfish's abdomen looks like a fan. The crayfish can curl this fan under itself very quickly. This helps the crayfish shoot backward. Then the crayfish is hard to catch.

22

On the outside of the crayfish's body is a shell. The shell is called an exoskeleton (ek-soh-SKEH-luh-tuhn). The exoskeleton is thin, but it is hard. It does not stretch. The exoskeleton is like armor. It protects the crayfish.

The exoskeleton protects a crayfish's soft insides.

Chapter 3

Crayfish hide during the day. They are active at night. What does a crayfish do at night?

Dinner Time

Adult crayfish are active at night. This is when they look for food. Crayfish do not like bright light. Sometimes they hide in shady spots during the day. Other times they hide under rocks or logs or in holes.

When crayfish hide, they go backwards into their hiding place. They keep their antennas pointing outward. Crayfish use their antennas and claws to tell if food or enemies are nearby.

Crayfish point their antennas and claws outward when they hide.

A crayfish uses its claws to hold food. Its claws also tear food into pieces. A crayfish's walking legs grab food and cut it up. Some of a crayfish's mouthparts hold food. Some crush food into small pieces. Crayfish also use their mouthparts to put food into their mouth.

Crayfish can grab and hold food with their claws.

A crayfish uses its walking legs to help it carry food.

Crayfish will eat almost anything. They often eat insects, small fish, and tadpoles. They also eat fish eggs, worms, and other crayfish!

A worm makes a good meal for a crayfish.

A favorite food for crayfish is snails. When crayfish eat snails, they even eat the shells. Eating snail shells helps make a crayfish's exoskeleton strong.

Crayfish eat plants as well as animals.

Some crayfish eat mostly plants. They eat plants that grow in the water.

Some crayfish eat dead animals and plants. These crayfish help keep lakes and rivers clean.

Chapter 4

A crayfish's eggs look like tiny black balls. What is a female crayfish called when she has eggs on her abdomen?

Growing Up

Most crayfish start families in the spring. Crayfish lay eggs. A female crayfish is very busy before she lays her eggs. First, she uses the tips of her walking legs to clean the bottom of her abdomen. Then she covers her abdomen and swimmerets with a type of glue. She makes this glue in her body.

This female crayfish is in berry. She curls her tail to protect her eggs.

Then the female lies on her back. She bends the end of her abdomen forward to lay her eggs. She covers her abdomen with eggs. The eggs grow stalks that attach to her swimmerets. A female with eggs on her abdomen is in berry. We say she is in berry because her eggs look like tiny berries.

The crayfish uses her swimmerets to move water over her eggs. This gives the eggs oxygen.

Different species of crayfish lay different numbers of eggs. The number of eggs depends on the size of the crayfish. Small crayfish lay fewer eggs than large crayfish lay. Some species of crayfish lay 10 eggs at a time. Other species lay as many as 800 eggs at a time.

Most female crayfish lay eggs two or three times in their life.

A crayfish who is in berry usually goes to a hiding place. There she waits for her eggs to hatch.

Crayfish eggs hatch when the water becomes warm. The eggs of some species of crayfish hatch in 2 weeks. The eggs of other crayfish take up to 20 weeks to hatch.

Mother crayfish take good care of their babies.

When the babies hatch, they are called first instars. First instars look like adult crayfish, but they are tiny. First instars stay attached to their mother.

The babies grow bigger. But their hard exoskeleton does not grow. The babies must shed their exoskeleton. This is called molting. For a few days, the babies' new exoskeleton is soft. Then it becomes hard.

Crayfish molt often when they are young. But they also molt when they are adults. An adult crayfish has shed this exoskeleton.

Shed exoskeletons are sometimes eaten by animals.

After molting, the baby crayfish are called second instars. They are no longer attached to their mother. But they hold on to her for 4 to 12 days. They molt again. Then they are called third instars.

This youngster has left its mother.

Soon the instars leave their mother. They go to live alone. The young crayfish look for food during the day and during the night. They grow quickly. By fall, the young crayfish have molted six to eight times. Then they are adults.

Chapter 5

Raccoons make crunchy meals of crayfish. What other animals eat crayfish?

Enemies

Crayfish have many animal enemies. Crayfish are food for fish, alligators, and raccoons. They are also food for otters, birds, and frogs.

It is easy for animals to eat crayfish just after they have molted. Right after molting, a crayfish's exoskeleton is very soft. But some animals can eat crayfish even when their exoskeleton is hard.

A fish has eaten part of this crayfish.

Crayfish have human enemies too. Some people use crayfish as bait for fishing. In some parts of the world, many people eat crayfish.

Humans catch crayfish. We use crayfish for food and for bait.

Crayfish protect themselves by trying to look big and scary.

Crayfish have ways to avoid danger. They usually scuttle away from an enemy. But sometimes they use their claws to try to scare enemies. Crayfish open and shut their claws. They wave them around. This way, crayfish look bigger than they are. If an enemy doesn't go away, crayfish use their claws to fight.

Crayfish often fight one another.

Sometimes an enemy grabs a crayfish by its leg. The crayfish has an amazing way to escape. It lets its leg fall off. The leg will grow back later.

Some people do not like crayfish. This is because some crayfish eat plants that humans grow. And sometimes crayfish dig through dirt dams that humans build. This lets out the water that the dams were supposed to keep in.

But crayfish are important. They are food for many animals. And crayfish help to keep lakes and rivers clean.

It is easy to find crayfish in freshwater. Some species even live in water that is slightly salty.

On Sharing a Book

As you know, adults greatly influence a child's attitude toward reading. When a child sees you read, or when you share a book with a child, you're sending a message that reading is important. Show the child that reading a book together is important to you. Find a comfortable, quiet place. Turn off the television and limit other distractions, such as telephone calls.

Be prepared to start slowly. Take turns reading parts of this book. Stop and talk about what you're reading. Talk about the photographs. You may find that much of the shared time is spent discussing just a few pages. This discussion time is valuable for both of you, so don't move through the book too quickly. If the child begins to lose interest, stop reading. Continue sharing the book at another time. When you do pick up the book again, be sure to revisit the parts you have already read. Most importantly, enjoy the book!

Be a Vocabulary Detective

You will find a word list on page 5. Words selected for this list are important to the understanding of the topic of this book. Encourage the child to be a word detective and search for the words as you read the book together. Talk about what the words mean and how they are used in the sentence. Do any of these words have more than one meaning? You will find these words defined in a glossary on page 46.

What about Questions?

Use questions to make sure the child understands the information in this book. Here are some suggestions:

> What did this paragraph tell us? What does this picture show? What do you think we'll learn about next? Could a crayfish live near your home? How do crayfish move? How do crayfish use their claws? Why must a crayfish shed its skin as it grows? What dangers must crayfish watch out for? What happens to a crayfish if one of its legs is broken off? What is your favorite part of the book? Why?

If the child has questions, don't hesitate to respond with questions of your own such as: What do *you* think? Why? What is it that you don't know? If the child can't remember certain facts, turn to the index.

Introducing the Index

The index is an important learning tool. It helps readers get information quickly without searching throughout the whole book. Turn to the index on page 47. Choose an entry, such as *eating*, and ask the child to use the index to find out what a crayfish eats. Repeat this exercise with as many entries as you like. Ask the child to point out the differences between an index and a glossary. (The index helps readers find information quickly, while the glossary tells readers what words mean.)

Where in the World?

Many plants and animals found in the Early Bird Nature Books series live in parts of the world other than the United States. Encourage the child to find the places mentioned in this book on a world map or globe. Take time to talk about climate, terrain, and how you might live in such places.

All the World in Metric!

Although our monetary system is in metric units (based on multiples of 10), the United States is one of the few countries in the world that does not use the metric system of measurement. Here are some conversion activities you and the child can do using a calculator:

WHEN YOU KNOW:	MULTIPLY BY:	TO FIND:
feet	0.3048	meters
inches	2.54	centimeters
gallons	3.787	liters
pounds	0.454	kilograms

Activities

Make up a story about a crayfish. Be sure to include information from this book. Draw or paint pictures to illustrate your story.

Watch a crayfish at the edge of a nearby lake or stream. What color is it? Find the crayfish's cephalothorax and abdomen. What other body parts can you identify? Is the crayfish using its legs or tail fan to move?

Visit a zoo to see crayfish and other crustaceans. How are crayfish similar to other kinds of crustaceans in the zoo and how are they different?

Glossary

abdomen (AB-duh-muhn)—the back part of a crayfish's body

antennas (an-TEH-nuhz)—the feelers on a crayfish's head. Crayfish use their antennas to feel, taste, and smell.

cephalothorax (seh-fuh-luh-THOR-ax)—the front part of a crayfish's body

compound eyes—eyes made up of many small eyes

crustaceans (cruhs-TAY-shuhnz)—animals with a hard outer shell, such as crayfish, lobsters, crabs, and shrimp

exoskeleton (ek-soh-SKEH-luh-tuhn)—a crayfish's hard outer shell

gills—the parts of a crayfish's body used for breathing

in berry—a female crayfish with eggs glued to her body

instars—baby crayfish

molting—shedding an old shell to make way for a new shell

oxygen (AHK-sih-juhn)—a gas most creatures need to live

swimmerets—leglike parts under a crayfish's abdomen

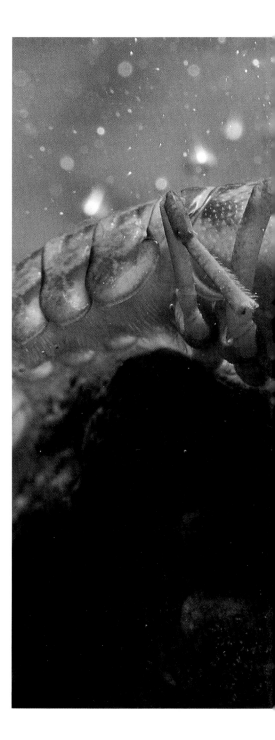

Index

Pages listed in **bold** type refer to photographs.

About the Author

Phyllis W. Grimm studied zoology in college, eventually earning a Ph.D. in botany and biochemistry in 1953. She interrupted her studies for two years, driving trucks in the U. S. Marine Corps Women's Reserve during World War II. She has worked in a number of jobs since receiving her degrees. Since retiring, she has done volunteer work presenting science programs in schools and has been involved in many environmental programs. She loves working with children and spending time with her grandchildren. She lives in Amery, Wisconsin.

About the Photographer

Jerry Boucher lives with his wife in rural Amery, Wisconsin. He has worked for over 30 years in photography, advertising, and graphic arts. His company, Schoolhouse Productions, does commercial photography, graphic design, and tourism brochures. The father of three sons, Boucher is also involved with Kinship, a teen photo group, and several arts organizations. He also teaches photography and drawing. His Lerner Publishing Group books include *Fire Truck Nuts and Bolts, Powerhouse, Flush!,* and *Rats.*